Dutch Waters

Dutch Waters

MASTERING THE FLOW OF CLIMATE RESILIENCE

Anurag Anurag

Contents

1

Introduction to Dutch Water Management

The Netherlands is a country uniquely defined by its relationship with water, with about a third of its land below sea level and another third at risk of flooding. This geographical vulnerability has necessitated a long history of sophisticated hydrological engineering, manifested in a landscape dotted with polders—tracts of land enclosed by dikes and drained by pumps—and extensive canal systems that regulate water levels. Historical disasters like the St. Elizabeth's flood of 1421 and the North Sea Flood of 1953, which resulted in significant loss of life and land, have profoundly shaped the national approach to water management, emphasizing the need for advanced, proactive measures.

In Dutch culture, managing water transcends engineering; it is a cornerstone of national identity, driven by a spirit of innovation, collaboration, and sustainability. The country is a recognized global leader in hydraulic engineering, a reputation upheld by institutions like the Rijkswaterstaat and various water boards responsible for implementing comprehensive water strategies. Contemporary water management practices in the Netherlands advocate for a symbiotic relationship with water, integrating safety and sustainability into both urban and rural planning. The education system reinforces this ethos, equipping future generations with the skills necessary for advanced hydrological and environmental management, thus ensuring the continuation and evolution of Dutch expertise in mastering water.

This deep-rooted expertise in managing water is not only about preventing disasters but also about embracing water as an ally. The Dutch philosophy of "living with water" has led to innovative projects such as the Room for the River program, which adjusts river courses to safely accommodate more water flow, and the Sand Motor, a dynamic method of coastal management that uses natural processes to replenish beaches and protect the shoreline. These projects demonstrate a holistic approach to environmental challenges, where the forces of nature are integrated into the solutions rather than merely countered.

The integration of water management into daily life is also visible in the Dutch landscape, with features like water plazas in urban areas that double as recreational spaces while providing flood relief when necessary. This blend of functionality and aesthetics is a hallmark of Dutch design, showing that infrastructure can enhance community living while serving critical environmental functions.

Furthermore, the Netherlands exports its water management expertise globally, assisting other nations in developing flood mitigation and water management strategies. This international collaboration not only highlights the Netherlands as a leader in the field but also emphasizes the global importance of innovative water management strategies in the face of climate change and rising sea levels.

By combining tradition with innovation, the Dutch have created a model of water management that balances the needs of safety, sustainability, and societal well-being, making it a source of national pride and an example for the world.

The success of Dutch water management extends beyond technical and infrastructural achievements; it also includes a strong public engagement and education component. Awareness and active participation in water management are cultivated from an early age through school programs and public initiatives that emphasize the importance of sustainability and proactive environmental stewardship. This public involvement ensures that water management is viewed not only as a governmental responsibility but as a shared societal duty.

Looking forward, the challenges posed by climate change will necessitate even more innovative and adaptable water management strategies. The Netherlands is continually testing and implementing new technologies, such as advanced computer modeling for flood prediction and innovative water purification techniques that enhance both water quality and availability. These developments are not just about keeping the water at bay but about creating resilient systems that can adapt to and absorb shocks, whether they come from natural forces or human impacts.

As a testament to its forward-thinking approach, The Netherlands hosts numerous international conferences and collaborates on global water management research projects, positioning itself as a knowledge hub for water-related challenges. Through these efforts, Dutch expertise is helping to shape a global response to water management that is adaptable, resilient, and sustainable.

In conclusion, the Dutch model of water management exemplifies how deeply intertwined human ingenuity and nature can be. It offers a beacon of hope and a wealth of lessons for countries worldwide facing the increasing threats of flooding and water scarcity driven by climate change. As we continue to explore Dutch strategies and innovations in subsequent chapters, it becomes clear that the path they chart not only secures their own future but also lights the way for global strategies in living with water.

2

~

History of Dutch Flood Control

The history of Dutch flood control begins with the early techniques that were crucial for survival in a landscape prone to flooding. Dykes, essential earthen barriers, were constructed to hold back the sea and protect the hinterlands. These were complemented by dams that controlled the flow of water through the intricate network of rivers and canals. Windmills played a pivotal role, not for milling grain as commonly thought, but for pumping water out of low-lying lands back into rivers and canals to maintain dry land and prevent waterlogging, an innovation that epitomized the Dutch mastery over their waterlogged environment.

The evolution of Dutch flood defenses took a significant leap forward with the construction of the Zuiderzee Works and the Delta Works.

Initiated in response to recurring devastating floods, the Zuiderzee Works transformed an inland sea into a freshwater lake and reclaimable land, drastically reducing the risk of flooding. This project not only showcased Dutch engineering prowess but also set the stage for the even more ambitious Delta Works. The Delta Works, a series of dams, sluices, locks, dykes, levees, and storm surge barriers, was designed to protect the southwestern provinces from the encroaching North Sea. Recognized as one of the modern wonders of the world, the Delta Works exemplifies the Dutch approach to comprehensive, long-term water management.

The catastrophic North Sea flood of 1953 was a turning point in Dutch water management, marking a shift from reactive to proactive flood defense strategies. The flood, which resulted in over 1,800 deaths and widespread destruction, underscored the need for robust flood defenses capable of withstanding extreme weather events. This disaster led to the accelerated completion of the Delta Works and spurred innovations in flood prediction and response systems. It instilled a lasting legacy of continuous improvement in flood management practices, emphasizing not only the construction of physical barriers but also the development of evacuation plans, public awareness campaigns, and environmental considerations to mitigate the impacts of future floods.

Through these historical milestones, Dutch flood control has evolved from simple methods of dykes, dams, and windmills to complex, integrated systems that protect the country from the ever-present threat of water. This history reflects a journey of innovation, learning from past challenges, and a steadfast commitment to safeguarding the land and its people against the forces of nature.

This relentless pursuit of innovation is exemplified in projects that integrate water management with environmental sustainability. The Dutch are pioneering techniques such as 'green dikes,' which incorporate natural vegetation to strengthen flood defenses while also supporting biodiversity. These projects illustrate how flood control can go hand in hand with environmental conservation, providing habitats for wildlife while protecting human settlements.

Additionally, the Dutch are exploring the use of advanced sensor technology and artificial intelligence to enhance their flood prediction and response systems. These technologies allow for real-time data collection and analysis, improving the accuracy of flood forecasts and enabling more effective deployment of resources in emergency situations. This integration of technology in water management not only boosts efficiency but also enhances the adaptability of the systems to face new challenges.

Public engagement remains a cornerstone of the Dutch approach to water management. Regular drills, educational programs, and community involvement initiatives ensure that citizens are not only aware of the risks but also actively participate in the safety measures. This widespread public involvement fosters a culture of preparedness and resilience, making water management a shared responsibility among all sectors of society.

As we move forward, the lessons learned from the Dutch history of flood control continue to influence global strategies for dealing with water-related challenges. The Netherlands' proactive and inclusive approach provides a blueprint for other nations grappling with similar issues, highlighting the importance of innovation, community involvement, and integrated planning in building resilient societies.

The enduring legacy of Dutch flood control is a narrative of adaptation and anticipation, reflecting a deep-seated respect for nature's power and a commitment to live in harmony with it. It serves as a reminder of what can be achieved when a nation unites behind a common cause, turning vulnerability into strength through the power of human creativity and collective effort.

In an era where climate change poses increasing threats to global stability, the Dutch model of water management offers invaluable insights. It exemplifies how a combination of historical knowledge, technological advancement, and community integration can form the backbone of effective environmental and disaster response strategies. This holistic approach is becoming increasingly essential as countries around the world face the dual challenges of urban expansion and extreme weather events.

Future advancements in Dutch water management are likely to focus on even greater integration of sustainable practices. Plans are underway to develop floating cities, utilize smart rainwater harvesting systems, and expand the use of aquifer storage and recovery. These innovations aim not only to mitigate the effects of flooding but also to make water resources more sustainable and resilient, aligning flood control with broader environmental goals.

Moreover, as global leaders in water management, the Netherlands continues to spearhead international efforts to build water-resilient communities. Through partnerships and collaborative projects, Dutch experts are helping to shape water management policies and infrastructures in vulnerable regions around the world. These efforts not only demonstrate global stewardship but also foster a shared understanding and action against the pervasive threat of water-related disasters.

The influence of Dutch water management practices extends beyond technical solutions; it has fostered a culture of resilience that emphasizes the importance of adaptive learning and collective action. As climate predictions become more severe, the Netherlands continues to refine its strategies, ensuring they are robust yet flexible enough to adapt to unforeseen circumstances. This adaptability is key to maintaining resilience in the face of the dynamic and evolving nature of global climate patterns.

In addition to structural and technological innovations, there is a growing emphasis on the socio-economic aspects of water management. This includes enhancing community resilience through education and participatory governance, ensuring that all layers of society are equipped to respond effectively to water-related challenges. The Dutch approach involves not just defending against water but using it as an opportunity to enhance social cohesion and community welfare.

This comprehensive strategy is evident in how the Netherlands plans for future scenarios with integrated water management that includes spatial planning, nature conservation, and urban development. By viewing water management as a cross-sectoral issue that touches on all aspects of society, the Dutch are able to create more sustainable and resilient communities.

Looking to the future, the Netherlands serves as a laboratory for innovation in water management. Experimentation with new materials and methods, such as bio-based dyke reinforcements and advanced monitoring systems, continues to push the boundaries of what is possible. These efforts are coupled with strong international dialogue and cooperation, sharing knowledge and learning from the experiences of others in similar situations.

As the world grapples with the increasing threats posed by climate change, the Dutch model offers a blueprint for turning vulnerability into a strength through comprehensive planning, community involvement, and innovative engineering. It stands as a testament to the power of human ingenuity and the potential for societies to not only survive but thrive by harnessing the challenges posed by their environments.

Building on the foundational practices of constructing dykes, dams, and windmills, the Netherlands has progressively embraced more sophisticated and integrative flood control strategies. The lessons from the 1953 North Sea flood catalyzed a transformative approach in Dutch water management, leading to the development of the Delta Works and the Zuiderzee Works—projects that not only embody the Dutch prowess in hydraulic engineering but also represent a shift towards creating resilient and multifunctional water infrastructures.

The Zuiderzee Works, initiated in the early 20th century, was a monumental undertaking that involved the damming and partial draining of the Zuiderzee, a saltwater inlet of the North Sea, to create new land, the IJsselmeer, and provide protection against sea flooding. This project significantly altered the landscape and helped secure agricultural and urban development areas from the threat of encroaching waters. It was a clear demonstration of the Netherlands' ability to reclaim land from the sea, a practice deeply rooted in Dutch history but executed on an unprecedented scale with the Zuiderzee Works.

Following the devastating floods of 1953, the Dutch government initiated the Delta Works, an extensive series of dams, sluices, locks, dykes, and storm surge barriers designed to prevent a similar catastrophe. This project was among the most sophisticated flood prevention systems in the world at the time and remains a critical component of the Netherlands' flood defenses. It includes structures like the Oosterscheldekering barrier, which not only protects against high tides and storm surges but also accommodates the ecological needs of the estuarine environment, highlighting the Dutch commitment to environmental sustainability alongside flood defense.

These projects have taught invaluable lessons about managing water in a densely populated, low-lying country. One key lesson is the importance of maintaining flexibility in flood defense systems to adapt to changing environmental conditions and technological advancements. Another lesson is the necessity of integrating water management with environmental and urban planning to ensure the sustainability of these projects. These insights have propelled the Netherlands to the forefront of global water management and continue to inform how the country and the world approach the challenges of rising sea levels and climate change.

As the Netherlands moves forward, the integration of these historical lessons with new technologies and strategies will be crucial in continuing to safeguard the country against future water-related challenges. This ongoing evolution in Dutch flood control strategy not only protects the physical and economic security of the nation but also enhances the quality of life for its residents, making it a model for effective water management around the globe.

3

~

The Maeslantkering: Engineering Against Extremes

The Maeslantkering, a part of the Netherlands' Delta Works, is a marvel of modern engineering and a testament to Dutch ingenuity in hydraulic engineering. This massive movable barrier was designed to protect the Rotterdam port and the surrounding areas from catastrophic flooding that could occur during severe North Sea storms. Its construction in the late 20th century marked a significant evolution in the approach to water management and flood control in the Netherlands.

The design of the Maeslantkering was driven by the need for a solution that could both protect against extreme water levels and preserve the busy maritime traffic to and from Rotterdam, one of the world's largest ports. The barrier comprises two gigantic arms, each as long as the Eiffel Tower is tall, that rest in docks on the riverbanks when open. These arms can close off the Nieuwe Waterweg, effectively preventing storm surges from reaching the inner land.

Operating the Maeslantkering involves a complex system of computers, sensors, and mechanics. The decision to activate the barrier is automated, determined by a computer system that analyzes tide and weather data to predict when water levels will rise to a point that threatens safety. Once a threat is detected, the gates, floating in their docks, are moved into position across the river. Hydraulic systems then fill with water, causing the gates to sink and form a watertight barrier. This operation is a delicate balance of engineering, hydrodynamics, and computer programming, designed to operate with minimal human intervention.

Within the broader context of the Delta Works, the Maeslantkering plays a crucial role. It is the final defense in a series of barriers designed to protect the southwestern part of the Netherlands, an area historically prone to severe flooding and where the consequences of a breach could be catastrophic not only locally but also nationally, considering the economic importance of Rotterdam. The Maeslantkering, therefore, not only serves as a physical barrier against flooding but also as a symbol of

Dutch resilience and technological advancement in the face of rising sea levels and climate change.

By protecting the economic heartland of the Netherlands, the Maeslantkering ensures that the country not only survives but thrives despite the environmental challenges posed by its unique geographical location. Its integration into the Delta Works highlights the interconnected nature of Dutch flood defenses, showcasing a comprehensive approach to managing the complex water systems that define the Dutch landscape.

The Maeslantkering's strategic importance is matched by its symbolic value, reflecting a commitment to sustainable and innovative engineering solutions. It embodies the proactive stance the Netherlands takes against the threats posed by climate change and sea-level rise, ensuring the safety and economic stability of the region. This barrier is not only a protective structure but also a benchmark for future engineering projects worldwide that seek to combine functionality with environmental consideration.

Moreover, the Maeslantkering has become a focal point for education and public awareness regarding flood management and climate adaptation. It serves as a learning center for engineers, policymakers, and the general public interested in understanding the complexities of water management. The facility includes a visitor center where people can learn about the history and operation of the Delta Works, emphasizing the continuous evolution of flood management strategies as environmental conditions change.

The ongoing maintenance and testing of the Maeslantkering are critical to its operation. Each year, a full deployment test is conducted to ensure that all mechanical systems function seamlessly. This test is not only a practical necessity but also a public demonstration of the robustness of Dutch flood defenses. It reassures the public and the international community of the Netherlands' preparedness to face extreme weather events.

As climate predictions become more severe, the role of the Maeslantkering within the Delta Works will likely evolve. Future enhancements may include integrating more advanced technologies for better prediction and response capabilities. The Netherlands continues to invest in research and development to keep its flood management technology at the forefront of innovation. This proactive approach ensures that structures like the Maeslantkering can adapt to future challenges, securing the safety and prosperity of the Dutch people in the long term.

Through its design, operation, and role within the broader Delta Works, the Maeslantkering stands as a pinnacle of Dutch engineering and a vital component of a comprehensive strategy aimed at combating the threats posed by the sea. It symbolizes the harmony between human ingenuity and nature's forces, serving as a model for sustainable development and resilience in water management across the globe.

The influence of the Maeslantkering extends beyond its immediate physical and operational capabilities; it serves as a critical case study for global flood defense initiatives. Nations vulnerable to sea-level rise and storm surges look to the Maeslantkering as a model of how to integrate large-scale engineering solutions with automated technology systems. Its success provides a blueprint for similar projects that aim to protect coastal cities and industrial areas while maintaining their operational functionalities.

Furthermore, the Maeslantkering's adaptability is a key feature, reflecting the need for infrastructural resilience in the face of climate variability. As global weather patterns become increasingly erratic, the flexibility offered by the Maeslantkering's design — allowing it to close only when necessary — minimizes impact on maritime traffic and local ecosystems. This approach underscores a broader philosophy in modern infrastructure development: the need for systems that are not only robust but also dynamically responsive to changing environmental conditions.

The integration of the Maeslantkering with the broader Delta Works also illustrates the importance of comprehensive planning in national safety strategies. It is a piece of a larger puzzle that includes not just physical barriers, but also policies on land use, environmental protection, and urban development. This holistic approach ensures that the safety measures are sustainable and that they enhance the ecological and social environments they are designed to protect.

In addition to its protective role, the Maeslantkering also contributes to the Netherlands' reputation as a hub of water management expertise. It attracts professionals and scholars from around the world who come to study its design and operation, facilitating an exchange of ideas that drives innovation in the sector. This international educational and professional exchange further solidifies the Netherlands' position as a leader in integrating engineering solutions with environmental and social governance.

As the world continues to face the challenges of climate change, the lessons learned from the Maeslantkering will undoubtedly influence future projects. Its legacy is not just in the protection it provides but in the forward-thinking approach it represents — an approach that balances human needs with those of the natural world, fostering resilience and sustainability for future generations. The Maeslantkering is more than just a barrier; it is a symbol of proactive adaptation and ingenuity, guiding the way toward more resilient and sustainable flood management solutions worldwide.

This proactive adaptation and ingenuity, exemplified by the Maeslantkering, spur ongoing developments not only in flood management but also in broader climate adaptation strategies. As such, the Maeslantkering is not a static monument but a dynamic, evolving system that reflects changing technological landscapes and environmental priorities. Its operational mechanisms and structural components are regularly reviewed and updated to incorporate the latest scientific insights and technological advancements, ensuring its efficacy in the face of new challenges.

The role of the Maeslantkering within the Delta Works also highlights the need for integrated systems that can communicate and operate in sync with one another. This is crucial in ensuring a coordinated response during emergency situations. The development of interconnected systems that can share real-time data and respond accordingly is a critical aspect of modern flood management strategies. It represents a shift from isolated infrastructure to networked systems that enhance resilience through collective operation.

Moreover, the environmental impact of such large-scale engineering projects is now a major consideration. The Maeslantkering's design takes into account not only the protection of human lives and property but also the preservation of local ecosystems. This dual focus is crucial in maintaining biodiversity and ecological health, which are integral to the region's environmental sustainability. Future modifications and upgrades to the Maeslantkering will likely emphasize reducing any negative ecological impacts while enhancing its functionality.

In a broader context, the Maeslantkering serves as a testament to the potential for human ingenuity to address seemingly insurmountable challenges through innovation. It is a beacon of hope and a source of inspiration for communities worldwide facing similar existential threats from water. By showcasing what can be achieved, the Maeslantkering encourages a global dialogue on sustainable, effective solutions to flood and climate-related challenges.

The legacy of the Maeslantkering, therefore, extends far beyond its immediate physical presence and function. It stands as a symbol of proactive global stewardship, a call to action for nations to embrace innovative and sustainable practices in water management. As climate change continues to pose significant global threats, the principles embodied by the Maeslantkering will remain critical in guiding future efforts to safe-guard vulnerable communities and ecosystems around the world.

As we continue to navigate the complexities of climate change, the Maeslantkering exemplifies how infrastructure can evolve to meet the needs of both safety and sustainability. It prompts a rethinking of how structures can serve multiple functions, not only protecting against immediate threats but also enhancing the long-term resilience of communities. This dual purpose is becoming increasingly crucial as urban areas expand and environmental conditions become more volatile.

The global impact of the Maeslantkering extends into the realm of policy and governance. It serves as a model for how governments can implement large-scale projects that require long-term vision and cross-sector collaboration. The success of such projects often hinges on the ability to forge strong partnerships between government bodies, private sector entities, research institutions, and local communities. These collaborations are essential in pooling resources, sharing knowledge, and driving innovation in ways that single entities cannot achieve alone.

Additionally, the Maeslantkering has catalyzed discussions on financing sustainable infrastructure. It demonstrates the importance of investing in high-quality, durable solutions that, while initially costly, provide immense long-term benefits in terms of risk reduction and economic stability. This approach challenges the conventional short-term cost focus, advocating for a broader perspective on budgeting for infrastructure that considers the total cost of ownership and the value of resilience.

Looking to the future, the principles learned from the Maeslantkering are being adapted to address other critical challenges, such as sustainable energy production and water resource management. These initiatives are part of a broader trend toward integrating infrastructure development with sustainable environmental practices, aiming to create systems that not only prevent disasters but also contribute positively to the environmental and social fabric of regions.

The legacy of the Maeslantkering is one of innovation, collaboration, and foresight. It embodies a commitment to tackling global challenges through smart, sustainable solutions that consider the well-being of both people and the planet. As the world faces increasing environmental threats, the insights gained from the Maeslantkering and the broader Delta Works project will continue to inspire and guide efforts to create a safer, more sustainable future.

This forward-thinking approach exemplified by the Maeslantkering has set a precedent for how nations around the world can approach their own infrastructural challenges, particularly in the realm of environmental sustainability. By integrating advanced engineering with ecological awareness, projects like the Maeslantkering show that it is possible to protect against natural disasters while also preserving the natural environment.

The knowledge and technology developed through the Maeslantkering project are being shared globally, forming part of international efforts to build resilient infrastructures capable of withstanding environmental stressors. This dissemination of knowledge underscores the importance of global cooperation in the face of shared challenges. It encourages an exchange of strategies and technologies that can be tailored to local needs while drawing on the extensive experience of projects like the Maeslantkering.

Moreover, the Maeslantkering's role in stimulating economic stability through protection of critical industrial areas like the Port of Rotterdam highlights the interconnectedness of infrastructure, economy, and environmental policy. This interdependence suggests that effective climate adaptation strategies can significantly contribute to economic resilience, providing a buffer against potential losses from climate-induced events.

As climate change continues to influence global policies and priorities, the lessons from the Maeslantkering will likely inform a new generation of infrastructure projects. These projects will increasingly need to balance functional demands with sustainability goals, ensuring that they not only mitigate immediate environmental risks but also contribute to long-term environmental health.

In conclusion, the Maeslantkering is more than a feat of engineering; it is a beacon of integrated thinking, where human safety, environmental sustainability, and economic vitality converge. Its ongoing legacy will inspire and inform future efforts to design and implement infrastructure that meets the complex challenges of the 21st century, making it a cornerstone example for future global resilience initiatives.

4

∾

Living with Water: Urban Water Management

In response to its unique challenges with water, The Netherlands has undergone a profound philosophical shift in its approach to urban water management, summarized by the mantra "living with water." This concept signifies more than just a defensive stance against water; it embodies a holistic integration of water into the fabric of urban life, acknowledging that water is both a perpetual threat and a vital resource.

In cities like Rotterdam, this philosophy has been brought to life through innovative urban design. The city's water plazas, for instance, are multifunctional public spaces that temporarily store rainwater during heavy showers, preventing flood damage. These plazas are designed not only for utility but also as aesthetic and social gathering places, showcasing how functionality and urban beauty can coexist. Floating architecture in Rotterdam and other parts of The Netherlands further exemplifies this approach. Communities and businesses are increasingly housed in structures that float on water, which not only mitigates flood risk but also creates dynamic living and working environments that adapt to rising and falling water levels.

The benefits of integrating water resilience into urban planning are manifold. Firstly, it enhances the safety and sustainability of cities in the face of climate change and rising sea levels, turning potential vulnerabilities into strengths. Water-resilient design also promotes biodiversity and environmental health by incorporating green spaces and natural water management solutions, which can improve air and water quality and provide habitats for urban wildlife. Economically, investing in water resilience can reduce future disaster recovery costs and increase property values by creating attractive and innovative urban environments. Socially, these spaces foster community interaction and recreational activities, enhancing the quality of life for residents.

This integrated approach not only protects but also enriches, positioning Dutch cities at the forefront of sustainable urban development.

It represents a model for cities worldwide, proving that adapting to water can bring about innovative solutions that enhance rather than hinder urban life. As this philosophy continues to evolve, it could redefine urban planning globally, demonstrating that living with water is not only about survival but about thriving in harmony with nature.

The Dutch approach to "living with water" extends beyond functional adaptations; it involves a significant cultural shift in how residents perceive and interact with their environment. Education plays a crucial role in this regard, as awareness and understanding of water management are integrated into the curriculum from an early age. Citizens are not only informed about the challenges and solutions regarding water management but are also actively involved in the process. This community engagement ensures that water resilience becomes a shared responsibility and a part of the collective consciousness.

Urban water management in the Netherlands also leverages technological innovation. Advanced sensor networks and real-time data analysis are increasingly used to monitor water levels, predict flooding events, and optimize the performance of water management infrastructure. These technologies enable cities to react quickly to potential threats and efficiently manage water resources, enhancing the adaptability of urban environments to changing conditions.

The "living with water" philosophy has spurred economic opportunities by fostering a new sector in water-resilient architecture and engineering. Dutch firms are now world leaders in this field, exporting their expertise to other flood-prone regions of the world. This not only contributes to global flood resilience but also promotes economic growth at home, creating jobs and stimulating innovation in related industries.

The integration of water resilience into urban planning also has a profound impact on the physical layout of cities. Traditional barriers and segregated water management systems are being reimagined as integrated, multifunctional landscapes that serve as parks, community spaces, and wildlife sanctuaries while providing essential flood protection. These spaces are designed to be floodable when necessary, ensuring that water can be accommodated without disrupting urban life.

As climate change continues to pose significant challenges to urban areas worldwide, the Dutch model of living with water offers valuable lessons on the benefits of integrating water management into the very fabric of urban planning. This approach not only mitigates the risks associated with water but also enhances the livability, sustainability, and resilience of urban environments. Looking ahead, this model provides a roadmap for other cities seeking to transform their relationship with water from one of vulnerability to one of vitality.

This progressive model also emphasizes the importance of urban green spaces in water management strategies. Parks, green roofs, and urban forests play crucial roles in absorbing rainwater, reducing runoff, and lowering flood risks. These green infrastructures serve the dual purpose of enhancing urban aesthetics and biodiversity while also acting as critical components of the city's water management system. Such integration of nature-based solutions aligns with global sustainability goals and promotes healthier urban living environments.

The success of these initiatives in the Netherlands has inspired international urban planning policies. Cities around the world are adopting similar strategies, recognizing that effective water management can significantly contribute to climate resilience. These adaptations are especially crucial in areas facing increased urbanization pressures and climate variability. By embracing the Dutch approach, cities can improve their adaptive capacity and ensure long-term sustainability.

The implementation of these water management solutions has fostered a new kind of urban creativity. Architects and planners are increasingly challenged to design multifunctional spaces that can handle water in innovative ways. This has led to a surge in creative urban design projects that not only address practical needs but also enhance the public realm.

Internationally, the Dutch philosophy of living with water has catalyzed a shift towards more proactive disaster preparedness and climate adaptation measures. By sharing their expertise through global forums, workshops, and consultancy, Dutch professionals are influencing urban development strategies worldwide. This collaboration extends beyond mere technology transfer; it involves a fundamental shift in how cities interact with their natural environments.

As this philosophy continues to evolve and spread, it reshapes the global discourse on urban resilience and sustainability. It presents an optimistic vision of the future where cities are not only safe from water-related threats but are also vibrant, dynamic, and harmonious ecosystems. This Dutch legacy offers a hopeful glimpse into a world where urban areas can flourish alongside their water environments, turning challenges into opportunities for growth and innovation.

The practical application of the Dutch "living with water" philosophy also includes the development of adaptive public infrastructure. In the Netherlands, roads, bridges, and public buildings are designed to accommodate water levels that fluctuate with changing weather patterns. For instance, roads and plazas in flood-prone areas are constructed with permeable materials that allow water to seep through, reducing surface runoff and enhancing groundwater recharge. This not only prevents flooding but also conserves water as a resource, illustrating a sustainable approach to urban infrastructure.

The Dutch approach integrates water storage solutions into urban landscapes. Examples include underground water storage systems that capture excess rainwater during peak downpours, which is later used for urban greening or even as emergency drinking water supplies. These systems are often invisibly integrated into urban parks, serving as recreational spaces during dry periods and as water storage during rainy seasons. This dual functionality maximizes space in densely populated cities while enhancing their resilience to water-related extremes.

The aesthetic aspect of water management is also significant in Dutch urban planning. Water features like canals, fountains, and ponds are not only functional for water management but are also central to the urban design, contributing to the cityscape's beauty and the inhabitants' quality of life. These features promote a connection to water in daily life, making it a visible and valuable part of the urban environment rather than a hidden or combative element.

In addition, the Dutch have pioneered the use of climate-adaptive buildings that respond dynamically to changes in weather and water levels. These buildings might feature floating foundations or water-resistant lower floors, and are equipped with technologies that adapt to different water conditions. Such innovations not only protect investments in property but also ensure continuity of service and safety for residents, showcasing a proactive approach to architectural design.

The societal benefits of this integrated water management strategy extend to public health and community well-being. Urban water features have been shown to reduce stress and increase feelings of well-being among city dwellers. Green spaces and water bodies contribute to cleaner air, lower city temperatures during heatwaves, and provide communal spaces for social interaction and physical activity.

Internationally, the Dutch model is enhancing global urban resilience by providing a template for others to follow, especially in regions newly facing the challenges of urban flooding due to climate change. Through workshops, collaborative projects, and consultancy, Dutch experts are helping cities worldwide to reimagine their relationship with water.

This deep integration of water management into every facet of urban planning and design not only prepares Dutch cities to face the future with confidence but also offers a blueprint for global cities aiming to transform their water challenges into sustainable opportunities. This holistic approach exemplifies how urban environments can become more resilient, sustainable, and livable by embracing and integrating their natural water systems into the heart of urban development.

Continuing with the Dutch model of integrated water management, a notable example is the Benthemplein Water Square in Rotterdam. This innovative project combines urban recreation with water storage capability. During dry weather, the square serves as a sports and community gathering area. When it rains, it transforms into a temporary water storage facility, helping to manage excess stormwater and reduce pressure on the city's drainage systems. This blend of functionality illustrates how urban design can adapt to both the climatic and social needs of a city.

Another example is the city of Amsterdam's initiative to develop floating neighborhoods such as IJburg. This area consists of several artificial islands where housing and infrastructure are built on platforms that float on water. These structures are designed to rise and fall with the water levels, providing resilience against flooding while also expanding the city's housing stock without sacrificing land resources. This project not only addresses housing needs but also integrates water resilience into the very foundation of urban expansion.

In terms of green infrastructure, the city of Utrecht has taken a proactive approach by integrating green roofs into its urban landscape. These roofs absorb rainwater, reduce heat in buildings, and decrease the load on sewer systems during heavy downpours. Utrecht's extensive green roofs program not only contributes to the city's sustainability goals but also enhances biodiversity, providing habitats for various species and improving urban air quality.

The Dutch have also pioneered the concept of "climate cafes" –

interactive workshops where citizens, scientists, policymakers, and urban planners come together to discuss and plan local adaptations to climate change. These forums are crucial in fostering public engagement and ownership of local water management solutions, ensuring that initiatives are well-supported and effective in meeting community needs.

The Netherlands also leads in deploying advanced technologies such as smart rainwater management systems. These systems use sensors and automated controls to manage the flow of rainwater through urban areas, optimizing the use of water resources and reducing the risk of flooding. For instance, in Eindhoven, smart sensors and controls allow the city to dynamically manage water levels in its ponds and waterways, ensuring that they serve both as beautiful features and effective flood management tools.

These examples showcase how the Dutch "living with water" philosophy is continuously evolving and adapting, creating a robust framework for urban water management that is being emulated worldwide. This approach not only mitigates the risks associated with water but turns these challenges into opportunities for enhancing urban environments. By viewing water as an asset rather than a threat, Dutch cities have fostered innovation, sustainability, and resilience, providing a model for cities globally as they prepare for the increasing uncertainties brought by climate change.

Expanding on the concept of water-resilient urban development, the Dutch model also emphasizes the reuse and recycling of water within urban systems. For instance, the city of Nijmegen has implemented a sophisticated water recycling system that treats wastewater to a high standard before reintroducing it into the local river system. This not only helps maintain the river's water level and quality during dry periods but also contributes to the overall health of the aquatic ecosystem.

In urban planning, the Dutch prioritize the concept of "blue-green" corridors. These are networks of natural waterways and green spaces that crisscross cities, providing essential ecological services such as habitat connectivity, enhanced biodiversity, and natural cooling of urban areas. These corridors also serve as major flood relief channels during high water events, absorbing and channeling excess water away from urban centers. An example of this is in the city of Groningen, where the restoration of the historic canals has improved water management while restoring natural habitats and providing recreational spaces for residents.

Dutch cities are increasingly turning to advanced digital modeling to predict and manage water-related challenges. For example, Delft University of Technology has developed sophisticated simulation models that cities can use to analyze flood risks and predict water flow through urban areas under various scenarios. These tools allow city planners to test and refine water management strategies in a virtual environment, reducing the risks and costs associated with physical interventions.

Another innovative approach is the development of water-aware urban furniture and infrastructure. In Rotterdam, for example, public benches double as water barriers and raised pedestrian pathways act as emergency flood routes. This kind of multipurpose infrastructure is not only cost-effective but also embeds resilience into the everyday urban fabric, ensuring that cities remain functional even during extreme weather events.

On a larger scale, the Dutch government's "Room for the River" program exemplifies a national commitment to adaptive water management. This initiative involves modifying the landscape to allow rivers more room to flood safely, thus reducing the risk to populated areas. It includes measures such as lowering floodplains, relocating dikes, and removing obstacles from river paths to enhance water flow. This program not only increases safety but also creates valuable new natural habitats and recreational areas.

Through these detailed examples and ongoing initiatives, the Netherlands demonstrates a comprehensive and forward-thinking approach to urban water management. This approach not only addresses the immediate needs of flood protection and water sustainability but also integrates these elements into the broader goals of enhancing urban livability and ecological health. The Dutch model continues to inspire cities worldwide, proving that effective water management is integral to the future of sustainable urban development.

The success of the Dutch approach to water management is further evidenced by their cutting-edge architectural designs that specifically cater to the challenges of living with water. For instance, in the Water District of Amsterdam, architects have designed floating homes, offices, and public buildings that are resilient to fluctuating water levels. These structures are not just functional; they are stylish and modern, appealing to a broad demographic and encouraging urban populations to embrace life on the water.

Dutch innovation extends into the realm of flood-resistant agriculture. The development of "floating farms" in areas like Rotterdam is a testament to the adaptive strategies being employed. These farms are built on floating platforms and are designed to operate fully independently of the land. They use rainwater harvesting systems and solar energy, minimizing their environmental impact while ensuring food security even during flood events. This approach is particularly revolutionary, combining urban farming with flood resilience to create a sustainable model for urban agriculture.

In transportation, the Netherlands is pioneering the integration of water management into roadway and railway systems. In areas prone to flooding, roads and rail lines are built with elevated sections or are equipped with flexible flood barriers that can be deployed during high water events. This ensures continuous connectivity and safety for transportation networks, vital for economic stability and emergency responses.

The educational impact of these initiatives cannot be understated. Dutch universities and research institutions are leaders in hydrological sciences and urban planning. They not only contribute to the national strategy on water management but also provide training and knowledge dissemination globally. Through academic programs, workshops, and consulting projects, Dutch experts are helping to build capacity for water management in countries facing similar challenges.

Community engagement is another pillar of the Dutch strategy. Regular public consultations, exhibitions, and interactive platforms ensure that citizens are not only aware of the water management projects but also have a say in them. This democratic approach fosters a sense of ownership and responsibility among residents, ensuring broad support for ongoing and future initiatives.

As Dutch water management strategies continue to evolve, they are increasingly being recognized as benchmarks for global best practices in resilience and sustainability. International delegations frequently visit the Netherlands to learn from its extensive experience, looking to adapt Dutch techniques to local conditions in their home countries. This global exchange enriches the Dutch approach, integrating new ideas and perspectives that enhance its effectiveness and applicability.

In conclusion, the Netherlands' comprehensive and integrated approach to living with water is a multifaceted strategy that encompasses architecture, agriculture, transportation, education, and community involvement. It exemplifies how a nation can not only adapt to life with water but thrive by turning environmental challenges into opportunities for innovation and growth. This model offers invaluable lessons for cities worldwide, demonstrating the potential for harmonious coexistence with nature in the age of climate change.

5

∾

Adaptation Strategies: From Theory to Practice

The Netherlands has effectively implemented flood management strategies that balance safety with environmental and social benefits. The Room for the River program is a notable example, designed to expand river floodplains and relocate dikes to increase water accommodation capacity during floods. This not only reduces flood risk but also enhances natural habitats and provides recreational opportunities for local communities.

Another significant project is the Maeslantkering storm surge barrier, part of the extensive Delta Works engineering project designed to protect the coastal regions of the Netherlands from flooding. This barrier is a marvel of modern engineering, with the ability to automatically close when storm surge levels threaten the safety of the Rotterdam area.

The country has also pioneered the use of multi-use spaces that function as both public amenities and flood management tools. For instance, water plazas in Rotterdam serve as playgrounds or sports fields during dry periods and as water storage basins during heavy rains. This innovative approach not only maximizes the utility of urban space but also engages the public in the city's water management efforts, making them an integral part of the landscape.

Public perception and community involvement are crucial to the success of these projects. The Dutch government ensures that these initiatives are transparent and inclusive, involving citizens in the planning process and keeping them informed about the benefits and operations. This inclusive approach fosters a sense of ownership and responsibility among residents, ensuring broad support and effective implementation of water management strategies.

These examples illustrate how the Netherlands has transformed its flood management practices from purely defensive measures to integrated, adaptive strategies that protect against flooding while improving urban spaces and engaging the community. Such strategies not only mitigate the effects of climate change but also provide a blueprint for sustainable urban development worldwide.

The innovative use of floating architecture in urban areas exemplifies another adaptation strategy. Cities like Amsterdam have embraced floating homes and offices, which are designed to rise and fall with water levels, thus mitigating flood risk while utilizing water as a living space. This adaptation not only addresses the challenges of urban expansion and sea-level rise but also adds a unique aesthetic and value to the cityscape.

Moreover, the Dutch approach includes sustainable drainage systems (SuDS) that are integrated into urban designs. These systems manage rainfall in a manner that reduces flooding and enhances the quality of urban environments. They include permeable pavements, green roofs, and rain gardens that absorb rainwater, decrease runoff, and decrease the burden on sewer systems. By managing water where it falls, SuDS contribute to a more sustainable and resilient urban water cycle.

Technological innovations also play a crucial role in enhancing the effectiveness of flood management strategies. Advanced predictive models and early warning systems utilize real-time data to forecast flood events with high accuracy, allowing for timely responses that minimize impacts. These technologies are supported by comprehensive data gathering networks, including sensors throughout the urban infrastructure that monitor water levels, rainfall intensity, and other relevant parameters.

Community engagement initiatives such as educational programs and participatory design workshops help to raise awareness about the importance of sustainable water management and encourage proactive community involvement. These programs are designed to inform residents about the impacts and benefits of adaptation strategies, fostering a culture of preparedness and cooperation. They also provide opportunities for residents to contribute their ideas and preferences, ensuring that the solutions implemented are well-suited to the needs and values of the community.

Through these multifaceted strategies, the Netherlands continues to lead by example in water management, showing how comprehensive planning and community involvement can turn the challenges of living with water into opportunities for creating more livable, adaptive, and resilient urban environments. This holistic approach not only secures the nation against the present and future risks associated with water but also enhances the social and environmental quality of its urban areas.

This proactive and integrated approach to urban water management has positioned the Netherlands as a global leader in adapting to the challenges posed by climate change and rising sea levels. Dutch cities are not only safeguarding against potential disasters but are also enhancing their livability and sustainability.

The ongoing commitment to innovation in water management has spurred further advancements. For instance, the development of climate-adaptive buildings which incorporate water-resilient construction materials and designs, such as elevated structures and waterproof basements, is becoming more common. These buildings are designed to cope with floods without significant damage, ensuring that urban life can continue uninterrupted even during adverse conditions.

Additionally, the Netherlands has expanded its focus to include the recovery and reuse of stormwater. Innovative projects are being implemented that capture and treat stormwater for non-potable uses such as irrigation and industrial processes. This not only reduces the strain on the city's freshwater resources but also minimizes the volume of water that must be managed during heavy rainfall events.

The success of these strategies has also inspired international cooperation and knowledge sharing. The Dutch government and private companies are heavily involved in water management projects worldwide, providing expertise and technology to help other nations develop their flood resilience strategies. This global engagement not only contributes to international water security but also opens up new economic avenues for Dutch companies specialized in water management technologies.

As urban areas around the world increasingly face the impacts of environmental change, the lessons learned from the Dutch experience offer valuable insights. These include the importance of adopting a flexible and multi-benefit approach to water management, the benefits of integrating nature-based solutions into urban planning, and the critical role of community involvement in ensuring the success and sustainability of adaptation measures.

By continuing to evolve and refine its strategies, the Netherlands sets a dynamic example of how countries can transform their relationship with water from one of vulnerability to one of strength, making the most of their water assets in a changing climate.

The Netherlands' approach to water management also emphasizes the importance of resilience planning that goes beyond merely preventing flooding. It includes the capability to recover efficiently and enhance systems based on lessons learned from each event. This iterative process ensures that each solution not only addresses current challenges but also anticipates future conditions, adapting to changes in climate, technology, and urban development patterns.

One of the key elements of this strategy is the incorporation of bioengineering techniques and ecological principles into traditional engineering projects. For instance, the use of living shorelines that include oyster reefs, mangroves, and salt marshes not only provides natural flood protection but also supports biodiversity and helps sequester carbon, contributing to broader environmental goals.

Moreover, the integration of smart city technologies plays a pivotal role in enhancing water resilience. Internet of Things (IoT) sensors, machine learning models, and big data analytics are employed to create highly responsive systems that optimize water usage and management in real-time. These technologies allow for more precise control over water infrastructure, such as pumps, gates, and reservoirs, ensuring that they operate at maximum efficiency and are responsive to the dynamic urban environment.

Furthermore, public-private partnerships have been crucial in advancing these initiatives. By collaborating with private sector innovators, research institutions, and non-governmental organizations, Dutch cities have been able to leverage a wide range of expertise and resources. These partnerships facilitate the development of cutting-edge solutions that are both economically viable and technologically advanced, ensuring that the financial and intellectual investment in water management generates substantial public benefit.

The comprehensive and collaborative approach taken by the Netherlands demonstrates that effective water management is not only about managing risks but also about enhancing the capacity of urban systems to deliver economic, social, and environmental benefits. This holistic view, when combined with a strong commitment to sustainability and community engagement, creates resilient urban landscapes prepared to face the uncertainties of the future.

As climate change continues to impact global communities, the Dutch model provides a blueprint for others to follow, highlighting that with thoughtful planning and cooperative effort, it is possible to turn water-related challenges into opportunities for growth and innovation. The Netherlands' ongoing journey in water management remains a leading example for cities worldwide, proving that proactive and integrated approaches can lead to sustainable and resilient urban development.

Continuing to push the boundaries of water management, the Netherlands has embraced advanced architectural and engineering solutions that serve as models for global best practices. For example, the city of Delft is home to experimental water projects like the WaterStraat, or Water Street, where new technologies and products for handling rainwater are tested in real-life urban settings. This outdoor laboratory allows companies, researchers, and students to experiment with paving materials, green roofs, and innovative water storage solutions under actual weather conditions. The initiative encourages creativity and rapid prototyping, speeding up the implementation of successful innovations into mainstream use.

In another pioneering effort, the city of Nijmegen has redeveloped its riverfront to enhance flood resilience through the creation of an elongated island in the River Waal. This project, part of the larger Room for the River program, involved moving the river dike inland and constructing a secondary channel to create room for the river to flood safely. This not only reduced flood risk but also revitalized the urban area, creating new recreational spaces, waterfront properties, and habitats for wildlife. The project successfully transformed a vulnerable flood area into a thriving district, showcasing how urban development can harmonize with river dynamics.

The concept of multi-functional landscapes is another area where the Netherlands excels. In Amsterdam, the Zuidas district, a major business and residential area, integrates water storage facilities beneath its buildings and public spaces. These facilities are part of a comprehensive water management system designed to cope with heavy rainfall and potential flooding while maintaining the area's functionality and aesthetic appeal. Such integrated solutions underscore the Dutch approach to making urban environments resilient in every aspect, not just in response to immediate threats but as a comprehensive, proactive strategy.

Furthermore, the Netherlands continues to lead in green infrastructure by integrating vegetation and water bodies into urban planning.

The city of Rotterdam, for instance, has developed green roofs and walls across many of its buildings. These features not only help manage stormwater but also reduce urban heat island effects, improve air quality, and provide residents with green, open spaces.

Each of these examples illustrates the Netherlands' forward-thinking approach to urban design and planning, where water management is seamlessly integrated into the fabric of the city. By prioritizing adaptability, innovation, and multi-functionality, Dutch cities ensure they are resilient, sustainable, and attractive places to live, even in the face of rising sea levels and increased rainfall due to climate change. This holistic approach offers valuable lessons for other cities worldwide, proving that integrating advanced water management strategies can significantly enhance urban resilience and sustainability.

Building on their extensive experience, Dutch cities are increasingly turning to adaptive public space designs that can change function based on weather conditions. For example, the Waterplein in Rotterdam is a public square that serves as a playground and sports area during dry weather and transforms into a rainwater collection and storage area during heavy rains. This innovative design not only maximizes urban space utilization but also serves as an educational tool, raising public awareness about the importance of water management in urban settings.

Additionally, the Dutch are pioneering in the use of adaptive road systems that can handle excessive water. In Eindhoven, a city known for its high-tech industry, smart roads equipped with porous surfaces and underground water storage systems can quickly absorb rainwater, preventing flooding and replenishing groundwater reserves simultaneously. These roads are also equipped with sensors that monitor water levels and can alert city managers in real time to potential flooding, allowing for swift responses.

The integration of art and culture into water management is another unique aspect of Dutch adaptation strategies. In several Dutch cities, public artworks double as water management tools. For instance, in the city of Leeuwarden, a large sculpture functions as a water storage tank during storms, blending urban beautification with practical functionality. These creative solutions not only make the cityscape more enjoyable but also embed the principles of sustainability and resilience into everyday life.

Research and development continue to play a crucial role in advancing these adaptive strategies. Dutch universities and research institutes collaborate closely with government, industry, and civic organizations to pioneer new technologies and methodologies for urban water management. This collaborative approach ensures that theoretical innovations are practically tested and refined before they are implemented, ensuring they are both effective and efficient.

As climate change increases the frequency and intensity of extreme weather events, the Netherlands' proactive and integrated approach provides a roadmap for other nations dealing with similar challenges. The Dutch model demonstrates that comprehensive urban water management can protect against future risks while enhancing the quality of urban life. By continuing to innovate and adapt, the Netherlands not only secures its own future but also contributes to the global conversation on sustainable urban development in an era of climate uncertainty.

The proactive stance of the Dutch also extends to governance and policy-making, where holistic and inclusive approaches are crucial. The Dutch government facilitates extensive cross-sector collaboration, involving private companies, local communities, non-profits, and international bodies in the planning and execution of water management projects. This approach ensures that all stakeholders have a voice in the process, increasing the projects' success rate and community buy-in. For instance, the "Water Boards," one of the oldest forms of local government in the Netherlands, are responsible for managing water levels, quality, and flood prevention. These boards are unique as they are democratic entities that directly involve the public in water governance through elections, making water management a deeply rooted civic responsibility.

In addition to governance, the Dutch focus heavily on scalability and replication of their water management strategies. They understand that solutions must not only work locally but also have the potential to be adapted elsewhere. The Amsterdam Rainproof initiative, for example, aims to make the city resilient to downpours. It encourages citizens, businesses, and the government to implement measures that can capture or delay the runoff from heavy rain. The strategies developed are shared widely, offering templates that can be adapted by other cities globally.

The Netherlands invests in international partnerships to combat global water management challenges. Dutch experts are often found working in flood-prone areas around the world, providing technical assistance and sharing knowledge gained from centuries of managing water. These efforts are part of the Netherlands' commitment to international cooperation on climate adaptation and sustainability.

The commitment to sustainability and innovation is also seen in the Dutch educational system. Universities and vocational schools offer specialized courses in water management, sustainability, and environmental sciences, creating a knowledgeable workforce that supports the nation's leadership in water management. This educational focus ensures ongoing innovation and a steady supply of skilled professionals to support sustainable practices.

The Dutch model's success demonstrates that effective water management is achievable with a combination of advanced engineering, robust policy, community involvement, and international cooperation. As the world faces increasing water-related challenges due to climate change, the Netherlands' approach offers a beacon of hope and a wealth of practical solutions that can be adapted to local contexts across the globe. This continuous cycle of learning, adapting, and sharing forms the backbone of what makes the Dutch water management system a paragon for others to emulate.

This ongoing cycle of innovation is further bolstered by the Netherlands' commitment to leveraging cutting-edge technologies. For instance, Dutch researchers and companies are at the forefront of developing smart water systems that use AI and IoT to optimize water usage and management. These systems not only enhance efficiency but also enable predictive maintenance, reducing the risk of breakdowns and ensuring continuous operation during critical periods. Such technologies are integrated into both urban and rural water systems, ensuring widespread benefits.

The Dutch also focus on future-proofing their infrastructure. Given the uncertainties associated with climate change, infrastructure is designed to be adaptable to a range of future scenarios. This includes building adjustable barriers that can be modified as sea levels rise or designing urban landscapes that can evolve as the population density changes. This forward-thinking approach not only safeguards against immediate water-related threats but also ensures that the infrastructure remains viable and effective in the long term.

In addition to physical infrastructure, the Dutch emphasize the importance of "soft" infrastructure such as community education, engagement, and behavior change. Programs are in place to educate citizens from a young age about the importance of water conservation and management. Public campaigns encourage residents to adopt rainwater harvesting systems and to reduce water usage. Such measures cultivate a culture of water consciousness, which is essential for the long-term sustainability of water resources.

The integration of nature-based solutions continues to be a key aspect of the Dutch strategy. Techniques such as creating wetlands, restoring natural floodplains, and using bio-engineering methods to strengthen coastal defenses not only provide effective flood protection but also enhance biodiversity and create recreational opportunities for the public. These nature-based approaches are often more cost-effective and sustainable than traditional engineering solutions, providing multiple ecological and social benefits.

The Dutch model of water management, with its blend of high-tech solutions, robust infrastructure planning, community involvement, and environmental stewardship, offers a comprehensive blueprint for other nations. As global environmental challenges intensify, the lessons from the Netherlands prove invaluable. They show that with the right mix of innovation, collaboration, and foresight, it is possible to turn the threat of water into an opportunity for enhancing urban resilience and sustainability. This holistic and adaptive approach serves not only the needs of the present but also prepares for the uncertainties of the future, ensuring that society can thrive in harmony with its water environment.

In conclusion, the Dutch approach to water management exemplifies a comprehensive and adaptive strategy that seamlessly integrates engineering excellence with ecological sensitivity and community engagement. Through a combination of advanced technology, robust infrastructure, and proactive policy, the Netherlands has established itself as a global leader in the field, turning its geographic vulnerability into a source of strength and innovation.

The success of the Dutch model lies in its holistic view of water management, which not only addresses immediate threats but also prepares for future challenges. This approach has been shaped by centuries of experience and is continuously refined through research, technological advancements, and community feedback. By prioritizing sustainability, resilience, and inclusivity, the Netherlands provides valuable lessons for other countries facing similar water-related challenges.

The Dutch commitment to international collaboration and knowledge sharing has allowed them to extend their influence well beyond their borders, helping to shape global water management strategies. This collaborative spirit is essential in an era where water issues are increasingly complex and transcend national boundaries.

As the world grapples with the pressing challenges of climate change, urbanization, and environmental degradation, the lessons from the Netherlands are more relevant than ever. By adopting and adapting the principles of the Dutch approach, other nations can enhance their resilience against water-related risks, ensuring safer, more sustainable, and thriving communities for future generations.

Thus, the Netherlands' journey in water management is not just a narrative of overcoming adversity; it is a beacon of innovative thinking and international cooperation, offering hope and direction for a water-wise world.

6

∽

Technological Innovations in Water Management

Recent decades have seen substantial advancements in water management technology, ranging from improved sensor systems to sophisticated modeling software that predicts water flow and flood risks with high accuracy. Innovations such as smart water grids and advanced filtration systems have revolutionized how cities handle both potable and waste water, enhancing efficiency and sustainability. For example, real-time data collection systems employed in these smart grids monitor water quality and usage across the urban landscape, allowing for dynamic adjustments that minimize waste and reduce costs.

Artificial intelligence (AI) and automation have become pivotal in enhancing the responsiveness and effectiveness of flood management systems. AI algorithms can analyze vast amounts of data from weather stations, satellites, and ground sensors to make accurate predictions about flood risk and activate preventive measures automatically. In the Netherlands, AI is used to operate complex mechanisms like storm surge barriers, which can decide autonomously to close in anticipation of rising water levels based on predictive analytics. This automation not only ensures rapid response times during potential flood events but also helps in resource allocation and emergency planning.

Looking ahead, future technologies in development for water management are focusing on even greater integration of AI and machine learning, with projects exploring autonomous robotic systems for dike inspection and repair, which could significantly reduce human risk and increase maintenance efficiency. Additionally, there is ongoing research into the use of nanotechnology for water purification, promising to deliver more effective methods for removing pollutants at the molecular level. These innovations highlight a trend towards more intelligent, resilient, and sustainable water management infrastructures that can adapt to changing environmental conditions and help mitigate the impacts of climate change.

The exploration of advanced materials also plays a critical role in the future of water management. Scientists are developing super-absorbent materials that can help in creating urban spaces that quickly absorb rainwater, reducing surface runoff and mitigating flood risks. These materials can be integrated into urban infrastructure such as sidewalks, playgrounds, and parking lots, making them multifunctional and contributing to more resilient cities.

In addition to material innovations, there is a growing focus on enhancing the energy efficiency of water management systems. New technologies are being developed that harness renewable energy sources to power water purification plants and pumping stations, reducing the carbon footprint associated with water management. Solar-powered desalination plants are one example, offering a sustainable solution to water scarcity issues, particularly in coastal regions.

The integration of geographic information systems (GIS) and remote sensing technology is also advancing the field of water management. These tools allow for the detailed mapping and analysis of water bodies, flood plains, and rainfall patterns, providing valuable data that can be used to improve water resource planning and management. By employing these technologies, water managers can better understand and predict the spatial and temporal patterns of water availability and use, leading to more informed decision-making.

The development of decentralized water systems is gaining traction as a way to enhance water resilience. These systems operate independently of larger municipal systems and can be crucial in times of disruption. For example, localized rainwater harvesting setups and on-site wastewater recycling systems can provide essential water supplies and sanitation solutions in emergency situations, reducing dependence on centralized systems that may be vulnerable to failure during extreme weather events.

The continuous advancement in technology and infrastructure is key to ensuring that water management strategies are not only effective but also sustainable and adaptive to the challenges posed by a rapidly changing global climate. As these technologies evolve, they offer the potential to significantly alter how communities manage and interact with water, ultimately leading to more secure and resilient water futures.

As technology continues to evolve, the push for smarter water management systems becomes increasingly crucial. Internet of Things (IoT) devices are being more widely implemented to create interconnected systems where water flow, pressure, quality, and usage can be monitored and managed in real-time across various sectors. This network of sensors and devices facilitates a holistic view of water systems, enabling precise control and optimization of water distribution and use, which is vital for conservation and efficiency.

Emerging technologies such as blockchain are also beginning to find applications in water management. By enabling secure and transparent transactions, blockchain technology can be used to manage water trading, track water allocations, and ensure compliance with water usage regulations. This can be particularly beneficial in regions where water scarcity requires stringent management and allocation strategies to ensure equitable distribution among users.

Artificial intelligence is set to further revolutionize water management by improving demand forecasting. Advanced predictive models powered by AI can analyze historical consumption data, weather forecasts, population growth trends, and other variables to predict future water needs with great accuracy. This capability allows water authorities to optimize their operations and maintenance schedules, reduce costs, and prepare more effectively for varying water demand scenarios.

Furthermore, the development of autonomous underwater vehicles (AUVs) and drones is enhancing the capabilities for exploring and monitoring aquatic environments. These tools can be deployed to inspect underwater infrastructure, survey flood-prone areas, or monitor the health of aquatic ecosystems without the need for human divers, providing safer and more cost-effective solutions for water management challenges.

The integration of these advanced technologies not only strengthens the infrastructure needed to manage water resources effectively but also enhances the ability of communities and nations to respond to environmental changes and water-related disasters with agility and informed precision. As we look to the future, the continued innovation in water management technology promises not only to enhance human well-being and environmental health but also to provide the tools needed to navigate the complexities of sustainable development in an increasingly water-stressed world.

The role of augmented and virtual reality (AR and VR) in water management is also beginning to unfold, offering new possibilities for training, planning, and public engagement. By using AR and VR, engineers and planners can visualize complex water systems and potential changes in real-time, enhancing decision-making processes and facilitating more effective communication among stakeholders. These technologies can simulate various flood scenarios or demonstrate the potential impacts of new infrastructure, helping to guide investments and community preparations for water-related events.

Adaptive learning algorithms are another frontier in water management, allowing systems to improve over time based on performance data. These algorithms can optimize water distribution, treatment processes, and maintenance schedules, continuously learning and adjusting to new data. As these systems become more sophisticated, they can anticipate problems before they occur, preventing failures and ensuring the stability of water supplies.

The convergence of digital twin technology with water management systems offers a powerful tool for optimizing the performance and resilience of water infrastructures. Digital twins create virtual replicas of physical systems, enabling managers to test scenarios and view the outcomes of various strategies without impacting the real world. This approach is invaluable for long-term planning and for responding to emergency situations, such as chemical spills or sudden infrastructure failures.

In the realm of public engagement and policy-making, the increasing availability of open data platforms and citizen science initiatives is democratizing water management. These platforms encourage public participation by allowing citizens to contribute data, monitor local water conditions, and engage with water management decisions. This inclusive approach not only fosters greater transparency but also enhances community resilience by empowering individuals with the knowledge and tools to act on water-related issues.

As these technological innovations continue to develop, they reinforce the necessity for robust cybersecurity measures to protect critical water management infrastructures. The integration of cybersecurity strategies ensures that systems remain secure from external threats and operational disruptions, which is crucial for maintaining public trust and the integrity of water services.

In conclusion, the future of water management is being shaped by a confluence of technologies that offer unprecedented opportunities to enhance efficiency, resilience, and inclusiveness in handling water resources. As the global community faces increasing challenges from climate change and population growth, these technological advancements provide essential tools for sustainable and adaptive water management strategies. The ongoing evolution in this field highlights the importance of continuous innovation, collaboration, and investment to harness the full potential of technology in securing a water-wise future for all.

7

Policy and Global Leadership

The Netherlands has long been recognized for its leadership in water management and climate adaptation strategies, stemming from its unique geographical challenges and extensive historical experience. This expertise is not only a matter of national policy but has also positioned the Netherlands as a global leader in sustainable water practices.

Dutch policies on climate adaptation and water management are comprehensive and integrated into national strategic planning. These policies emphasize not only the protection and sustainable use of water resources but also the necessity of adapting urban and rural landscapes to the realities of climate change. The approach is holistic, considering everything from agricultural practices to urban infrastructure, ensuring that all aspects of water management are sustainable and resilient.

The Netherlands' contributions to global water management practices are extensive. Dutch firms and government agencies are often involved in water projects worldwide, providing expertise and technology that help other nations develop flood defenses, manage river basins, and treat wastewater. This international collaboration is part of the Dutch commitment to sharing knowledge and fostering global responses to water-related challenges.

Furthermore, the Dutch government actively promotes international cooperation through various programs and initiatives. For example, the Netherlands Water Partnership (NWP) works as a comprehensive network of Dutch organizations and businesses focused on water management. It facilitates collaboration between the private sector, NGOs, and governmental agencies to address global water issues. The NWP not only helps in disseminating Dutch water management innovations but also assists in tailoring these innovations to fit local needs in countries facing water stress.

Dutch aid and collaboration extend beyond technology transfer; they also involve capacity building and governance training. Through partnerships with multilateral organizations such as the United Nations and the World Bank, the Netherlands provides funding and expertise to help developing countries implement effective water management strategies and policies. This aid reflects a deep understanding of the interconnected nature of global water issues and a commitment to solving them through cooperation and mutual development.

The impact of Dutch leadership in water management is also evident in its proactive stance on climate policy at international forums. The Netherlands has been a vocal advocate for more stringent global water and climate policies, pushing for increased investment in climate resilience and sustainable development worldwide. Through these efforts, the Netherlands not only helps to shape international environmental policy but also ensures that water management remains a central topic in discussions about global sustainability and climate adaptation.

The Dutch approach to policy-making in water management also includes significant investment in research and development. The government often collaborates with academic institutions and research centers to pioneer new techniques and technologies that can later be implemented worldwide. For example, the Deltares Institute in the Netherlands is a leading international research institute in water and subsurface management, providing innovative solutions to complex water challenges. These solutions range from flood risk assessments and smart infrastructure designs to ecosystem restoration projects. By investing in research, the Netherlands not only enhances its own capabilities but also contributes to the global knowledge base on water management.

International training programs and workshops are another crucial aspect of Dutch contributions to global water management. These programs are designed to train water professionals from around the world in the latest techniques and strategies in water management, disaster preparedness, and climate adaptation. By doing so, the Netherlands helps to build a more skilled workforce globally, capable of tackling water-related challenges in their home countries.

The Dutch also leverage their extensive experience in negotiating water treaties and managing transboundary water resources as a diplomatic tool, promoting cooperation and peace. They have been instrumental in facilitating discussions and mediating conflicts over water resources in various parts of the world. This diplomatic engagement not only helps resolve potential conflicts but also promotes integrated and cooperative approaches to water management, which are essential in regions where water resources span multiple political boundaries.

The Netherlands hosts numerous international conferences and forums on water management, attracting experts, policymakers, and activists from around the globe. These events serve as platforms for sharing best practices, forging new partnerships, and coordinating international responses to water crises. For instance, the Amsterdam International Water Week is a prominent event that brings together leaders in water management to discuss innovative solutions to water challenges.

The Netherlands' policy and global leadership in water management demonstrate a deep commitment to sustainability, innovation, and international cooperation. Through a blend of robust domestic policies, active global engagement, and a focus on education and research, the Netherlands continues to influence water management practices worldwide, helping to ensure that water resources are managed wisely and sustainably for future generations.

Expanding its role as a global water steward, the Netherlands actively participates in shaping international environmental protocols and agreements. The nation's expertise and proactive policy initiatives are crucial in global negotiations such as the Paris Agreement, where they advocate for comprehensive strategies to mitigate climate change impacts on water resources. Their contributions often emphasize the need for global resilience through improved water management and disaster risk reduction frameworks, which are essential for protecting vulnerable ecosystems and human populations.

The Dutch model also emphasizes the importance of integrated water resources management (IWRM), a process that promotes the coordinated development and management of water, land, and related resources to maximize economic and social welfare without compromising the sustainability of vital ecosystems. The Netherlands advocates for IWRM principles in international water management policies, encouraging countries to adopt holistic management practices that consider entire water basins, which can span multiple national boundaries.

Furthermore, the Dutch government supports numerous international development projects aimed at improving water management in developing countries. These projects often involve infrastructure development, such as constructing dams and irrigation systems, but also focus on improving water governance and institutional capacity. The goal is not only to provide immediate improvements in water access and quality but also to leave a lasting impact by establishing robust governance frameworks that can sustainably manage water resources long after the projects have concluded.

Innovation continues to be a hallmark of Dutch water management, with ongoing development in areas such as saline agriculture, which involves using saltwater for irrigation to grow crops in arid regions where freshwater is scarce. This technology has significant implications for food security in climate-vulnerable regions and represents a critical area of research and development where the Netherlands leads.

Additionally, the Netherlands fosters a vibrant startup ecosystem focused on water technology. This ecosystem is supported by government funding, research institutions, and a collaborative business environment that encourages innovation in areas like water-saving technologies, advanced desalination techniques, and efficient water recycling systems.

Startups often receive support to scale their operations internationally, further extending the Dutch influence on global water management practices.

The proactive and innovative strategies of the Netherlands in water management and international collaboration provide a blueprint for other nations to follow. By continuing to lead by example and supporting global efforts to address water challenges, the Netherlands plays a vital role in promoting sustainable water use and management worldwide, ensuring that future generations can enjoy the benefits of secure and resilient water resources.

The Netherlands' commitment to global water management is also showcased through their significant role in climate resilience initiatives. Recognizing that water management is intricately linked to climate adaptation, the Dutch government integrates its water management strategies with its climate policies. This ensures a comprehensive approach that addresses both mitigation and adaptation needs, reinforcing infrastructure and communities against the increasing unpredictability of climate impacts.

The Netherlands has been a driving force behind several multinational research collaborations aimed at addressing water scarcity and improving water efficiency. These collaborations often result in groundbreaking technological innovations, such as advanced membrane technologies for water purification and novel approaches to wastewater treatment that recover nutrients and generate energy, making the process more sustainable and less resource-intensive.

In its pursuit to lead global water diplomacy, the Netherlands also plays a pivotal role in transboundary water management. They work closely with nations sharing river basins to develop agreements that ensure equitable and sustainable water distribution, emphasizing the need for cooperative management structures that can adapt to changing hydrological realities due to climate change.

Dutch aid programs are crucial in promoting water security in developing regions. These programs focus on both emergency water supplies in response to crises and long-term investments in water infrastructure that foster development and stability in these regions. For example, Dutch-funded projects have helped build resilient agricultural practices in sub-Saharan Africa by introducing drip irrigation systems that conserve water while boosting crop yields.

The Netherlands also champions the cause of water and sanitation at major global forums, advocating for universal access to clean water and sanitation as fundamental human rights. Through its influence in international bodies like the United Nations, the Netherlands pushes for increased funding and attention to global water and sanitation challenges, aiming to achieve the Sustainable Development Goals related to water.

By continuing to innovate and advocate for integrated and sustainable water management, the Netherlands not only enhances its resilience but also contributes significantly to global efforts. Their integrated approach, combining technology, policy, and international cooperation, serves as a model for other countries looking to navigate the complex challenges of water management in an era of increasing environmental uncertainties.

The Dutch approach to water management is not just about responding to challenges but also about proactively shaping the future of global water sustainability. This vision is encapsulated in their leading role in initiatives like the High-Level Panel on Water, where they work alongside other nations to catalyze change in water policies at the global level. These efforts emphasize the integration of water into all aspects of sustainable development, from urban planning and energy usage to agriculture and environmental protection.

Recognizing the link between education and sustainable water management, the Netherlands invests heavily in educational programs that focus on water sciences. Universities and technical schools in the Netherlands are renowned for their water-related curricula, which attract students from around the world. These institutions not only provide cutting-edge research and development but also foster a new generation of water professionals equipped to handle both local and global water challenges.

The Netherlands also utilizes its expertise to support disaster risk reduction globally. Dutch flood management experts frequently collaborate with international agencies to provide consultancy and aid in constructing flood defense systems, especially in areas frequently affected by severe weather events. This practical support is complemented by Dutch involvement in global climate change discussions, advocating for robust, actionable strategies to mitigate the impact of extreme weather on vulnerable populations.

Innovation continues to be a driving force in the Dutch approach to water management. The country is a hub for water technology startups and incubators that translate scientific research into practical solutions. These businesses benefit from strong governmental support, which includes grants, subsidies, and facilitated access to international networks. Such support underscores the Netherlands' commitment to fostering innovation that can be shared globally to improve water resilience.

Furthermore, the Netherlands hosts numerous international conferences and summits focused on water issues, providing a platform for world leaders, experts, and activists to collaborate and share knowledge. These events not only spotlight the latest advancements in water management but also facilitate multinational agreements and initiatives that aim to address the pressing water challenges of our time.

Through a combination of technological innovation, policy leadership, and global collaboration, the Netherlands continues to strengthen its role as a global leader in water management. This proactive and comprehensive approach not only ensures the sustainability of Dutch water resources but also contributes significantly to solving some of the most pressing water-related issues facing the world today. As climate change and population growth continue to exacerbate water challenges, the Dutch model of integrated and innovative water management remains a beacon for nations striving to secure a sustainable water future.

The Netherlands also leverages its expertise in water management to foster international economic cooperation and development. By exporting water management technologies and offering consultancy services, Dutch companies are pivotal in building water resilience globally while stimulating the Netherlands' economy. This approach aligns with the Dutch policy of using water management as a strategic economic sector, recognizing its potential to drive innovation, sustainability, and international trade.

As part of their commitment to global water security, the Netherlands supports various initiatives under the United Nations' framework, actively participating in projects that address water and sanitation issues in underdeveloped countries. These initiatives not only provide immediate relief but also aim to establish sustainable practices for long-term water management, integrating community involvement and local governance to ensure effectiveness and sustainability.

Moreover, the Dutch influence extends into the realm of international law and policy, where they advocate for regulations that protect rivers, seas, and transboundary water resources. The Netherlands plays a critical role in shaping international treaties and agreements that aim to manage shared water resources equitably and sustainably, promoting peace and cooperation between nations.

In addition to these global initiatives, the Netherlands focuses on climate resilience within its borders and in European contexts. It actively participates in the European Union's strategies for water management, contributing to policies that tackle pollution, manage floods, and address the impacts of climate change on water bodies. This leadership role not only enhances regional water security but also sets standards that can be emulated by other regions.

The ongoing commitment of the Netherlands to advancing global water management is a testament to its belief in the interconnectedness of water issues and the need for collective action. By continuing to drive innovations, share knowledge, and foster international cooperation, the Netherlands not only cements its status as a leader in water management but also contributes to the global effort towards achieving water sustainability for all. This holistic approach ensures that water management remains at the forefront of global environmental, economic, and social agendas, helping to navigate the challenges of the 21st century.

In conclusion, the Netherlands' approach to water management exemplifies a profound commitment to innovation, sustainability, and global leadership. Through a combination of robust national policies, groundbreaking technological advancements, and extensive international collaboration, the Netherlands has not only secured its own water resilience but also significantly contributed to global water management practices.

Dutch policies on climate adaptation and water management are integrated and forward-thinking, ensuring that the nation remains at the cutting edge of managing its water resources in harmony with environmental and societal needs. Internationally, the Netherlands has become synonymous with excellence in water management, often called upon to share its expertise and lead initiatives that shape global water strategies.

The collaborations and aid provided to other countries underscore the Dutch dedication to fostering global water security. Through initiatives ranging from direct aid and infrastructure projects to participation in high-level policy-making forums, the Netherlands leverages its small size into a significant global impact, promoting sustainable water use and management around the world.

As we face increasing global challenges related to water—from scarcity and pollution to flood risks—the Dutch model offers valuable lessons and a framework for other nations to emulate. By continuing to innovate and collaborate, the Netherlands not only enhances its own resilience but also actively contributes to building a more sustainable and water-secure world. This leadership and commitment make the Netherlands a pivotal player in the ongoing global dialogue on water management and climate adaptation.

8

Challenges and Future Prospects

Despite the Netherlands' renowned expertise in water management, the country faces ongoing and emerging challenges that require continual adaptation and innovation. One of the primary concerns is the rising sea level, a direct consequence of global warming, which threatens to overwhelm existing coastal defenses. The low-lying nature of much of the Dutch landscape, with significant portions below sea level, makes this an especially acute issue.

The increase in extreme weather events such as heavier rainfalls and more intense drought periods poses a dual threat. Urban drainage systems, even in well-prepared cities like Amsterdam and Rotterdam, are sometimes pushed beyond their capacity during severe rainstorms, leading to urban flooding. Conversely, drought conditions can severely impact water supply, agricultural productivity, and natural ecosystems, necessitating more robust and flexible water resource management strategies.

Looking at the impacts of climate change on infrastructure, the aging of water-related infrastructure like dikes, dams, and sluices also presents significant challenges. Many of these structures were built to standards that may not withstand the current predictions for climate change impacts. The potential for infrastructure failure poses a risk not only to public safety but also to the economy, particularly in a country where much of the GDP is generated below sea level.

In response to these challenges, the Netherlands is planning a series of future projects and adaptations. One key area of focus is the strengthening and heightening of sea defenses. Projects such as the Delta Program are being continuously updated to incorporate the latest scientific predictions for sea-level rise. This program aims to secure all sea defenses against a 1 in 10,000-year storm event, a standard that is among the highest in the world.

The Dutch are pioneers in implementing "building with nature" projects. These initiatives use natural processes and materials to create water management solutions that are both effective and environmentally sustainable. For example, the Sand Motor project involves creating a massive artificial sandbank offshore which naturally reinforces the coastline as the tides distribute the sand.

The Netherlands is also investing heavily in research and technology to better predict weather events and their impacts on water systems. Advances in computational power and data analytics are enabling more accurate models and simulations of flood scenarios, which in turn inform infrastructure design and emergency response strategies.

The Dutch are extending their expertise and collaborative approach globally, as part of their vision for water management. They actively engage in international water management projects, providing expertise and support to improve global water resilience. This not only helps other nations but also ensures that Dutch researchers and engineers remain at the forefront of global water management practices.

While the challenges facing Dutch water management are significant and evolving, the country's commitment to innovation, research, and international collaboration positions it well to continue its leadership role in this critical field. The future projects and ongoing adaptations reflect a proactive approach to ensuring that the Netherlands not only survives but thrives in the face of global water-related challenges.

The Dutch commitment to sustainability and resilience is further illustrated in their proactive approach to urban water management. Cities across the Netherlands are increasingly incorporating green infra-structure into their landscapes, such as permeable pavements, green roofs, and expanded wetlands. These features not only help manage stormwater and reduce flood risk but also contribute to biodiversity and improve the urban microclimate, making cities more livable and environmentally friendly.

In terms of water supply, the Netherlands is exploring innovative ways to ensure a reliable and clean water supply in the face of increasing salinity of rivers and groundwater due to sea-level rise. Projects are under-way to enhance freshwater storage capacity, utilizing deep aquifers and creating new water reservoirs that can serve as buffers during periods of drought or saltwater intrusion. These measures are essential to maintain-ing the balance of freshwater ecosystems and ensuring the sustainability of agricultural and industrial water uses.

The Dutch are also at the forefront of developing circular water economy models, aiming to minimize waste and maximize reuse within their water systems. By treating wastewater to high standards, they are able to recycle it for various purposes, including industrial processes, irrigation, and even as drinking water after sufficient purification. This not only conserves precious water resources but also reduces environmental pollution.

To address the future impacts of climate change, the Netherlands is engaging in scenario planning and adaptive management strategies. These strategies are designed to be flexible and scalable, allowing adjustments based on actual changes in climate conditions and technological advancements. This adaptive approach ensures that water management practices can remain robust and effective even under uncertain future scenarios.

Internationally, the Netherlands continues to share its knowledge and experience, participating in global water management initiatives and helping other countries develop their own adaptive strategies. This cooperation is carried out through various international platforms and involves a range of activities from joint research projects and technology transfers to policy advising and capacity building.

Overall, the future prospects for Dutch water management are shaped by a deep understanding of the challenges ahead and a clear vision for innovation and collaboration. By continuing to lead in technological advancements, policy development, and international cooperation, the Netherlands is well-positioned to face the evolving global water management landscape, setting a benchmark for others to follow in the pursuit of sustainable and resilient water practices.

As climate change continues to alter hydrological cycles globally, the Netherlands is intensifying its focus on climate-proofing its infrastructure. This involves upgrading urban systems to handle extreme weather events and incorporating flexible design standards that can adapt over time as climate predictions update. For instance, water systems are being designed with modular components that can be easily expanded or modified to accommodate increased capacities as needed.

This forward-thinking approach ensures that infrastructure investment remains protective and productive long into the future.

The Netherlands is also expanding its research into emerging technologies such as artificial intelligence (AI) and the Internet of Things (IoT) to enhance water management. AI is being integrated into predictive models to improve the accuracy of flood and drought forecasts, allowing for more precise water management and disaster response. IoT devices are being deployed throughout the water management infrastructure to provide real-time data on water levels, quality, and infrastructure integrity. This data-rich approach facilitates a more responsive and informed management system that can dynamically adapt to changing conditions.

Recognizing the importance of a holistic approach to water challenges, the Netherlands advocates for and practices integrated water resource management (IWRM). This strategy considers the entire water cycle from source to sea, encompassing all uses and users of water resources in a unified framework. This comprehensive approach ensures that all aspects of water management, from environmental to economic to social, are considered and balanced in decision-making processes.

Furthermore, the Dutch are pioneers in promoting water consciousness among the public and stakeholders. Educational campaigns, public participation in water-related decision-making, and transparency in water management operations are central elements of the Dutch strategy. This societal engagement fosters a culture of water stewardship, ensuring that all segments of the population understand the challenges and contribute to the solutions.

On the global stage, the Netherlands continues to offer its expertise in areas particularly vulnerable to climate impacts. Through partnerships and aid programs, Dutch experts are helping to design and implement water management solutions in regions facing severe water scarcity, flooding, and pollution. These efforts are part of the Netherlands' commitment to global water security and sustainability, recognizing that water issues are inherently global and interconnected.

The Dutch approach to water management, characterized by innovation, integration, and international cooperation, not only prepares the Netherlands for future challenges but also contributes to global efforts to manage water sustainably. As the world confronts more severe and frequent water-related challenges, the Dutch model provides valuable insights and practical solutions that can be adapted to diverse contexts, helping to secure a water-resilient future for all.

Looking forward, the Netherlands is investing in the development of next-generation water technologies that could redefine the future of water management. This includes advancements in desalination techniques, where research is focused on reducing energy consumption and increasing efficiency through novel methods such as forward osmosis and biomimetic membranes. These technologies aim to make desalination more viable and sustainable, potentially solving water scarcity issues in coastal regions worldwide.

The Dutch are exploring the potential of smart water networks that use advanced analytics to optimize water distribution and use. These networks could automatically adjust flow rates, detect leaks, and manage water allocations in real-time, ensuring optimal efficiency across various sectors. Such systems would not only conserve water but also significantly reduce operational costs and environmental impacts associated with water distribution.

The Netherlands is also a leader in fostering blue-green infrastructure, which integrates water management solutions with urban development to create sustainable and resilient cities. This includes the creation of urban wetlands, restoration of natural water courses, and the development of green roofs and walls that absorb rainwater. These initiatives not only help manage stormwater and reduce flood risks but also enhance urban biodiversity and provide residents with green recreational spaces.

To support these technological and infrastructural advancements, the Dutch government is also focused on policy innovation. This includes developing regulatory frameworks that encourage the adoption of sustainable water practices, such as mandating water-neutral building developments or incentivizing the reuse of greywater. Policies are also being crafted to support the transition towards a circular water economy, where waste water is seen as a resource rather than a disposal problem.

Internationally, the Netherlands is expanding its role as a thought leader in water management through the Water as Leverage program, which explores how water-related challenges can catalyze broader social and economic improvements in vulnerable regions. This program exemplifies how Dutch expertise is not only about exporting solutions but also about empowering local communities to innovate and adapt based on their specific circumstances.

As the Netherlands continues to push the boundaries of what is possible in water management, it remains committed to sharing its knowledge and innovations with the world. This global outlook not only enhances the capabilities of other nations to manage their water resources effectively but also builds a more connected and cooperative international community ready to tackle the complex water challenges of the future. Through ongoing research, policy development, and international collaboration, the Netherlands is helping to pave the way for a more sustainable and water-secure world.

The Netherlands is also keenly aware of the need to enhance resilience against the consequences of climate change on the water cycle. As such, there's an increased focus on developing resilient agricultural practices that require less water and are more tolerant of water quality fluctuations. Innovations in agricultural technology, including precision irrigation systems that deliver water directly to the root zone of plants, are being implemented to minimize water waste while maximizing crop yields. This approach not only conserves water but also supports sustainable food production in the face of changing climate conditions.

The Dutch water sector is harnessing the power of big data and analytics to improve decision-making and resource allocation. By collecting and analyzing data from various sources, including satellite imagery and sensor networks, water managers can gain a comprehensive understanding of water systems. This information is crucial for predicting water demand, planning resource distribution, and implementing conservation measures effectively. Advanced data analytics also enable the detection of patterns and trends that can inform long-term water management strategies, ensuring that resources are managed sustainably and efficiently.

To bolster these technological and analytical advancements, the Netherlands advocates for strong global governance on water issues. The Dutch government actively participates in international policy discussions and treaties related to water management, advocating for standards and practices that promote sustainable and equitable water use worldwide. Through these diplomatic efforts, the Netherlands not only contributes to shaping global water policy but also ensures that it aligns with the highest standards of sustainability and fairness.

Building on its foundation of innovation and cooperation, the Netherlands is also spearheading initiatives to address the energy-water nexus, recognizing the interdependencies between water management and energy consumption. Efforts to integrate renewable energy sources into water infrastructure are gaining momentum. For instance, solar and wind energy are increasingly being used to power desalination plants and wastewater treatment facilities, reducing the carbon footprint associated with water management operations while ensuring energy security.

The Dutch are exploring the use of advanced materials and

technologies to further enhance the efficiency and resilience of their water systems. For example, the development of ultra-permeable concrete and other novel materials that can quickly absorb and release water is underway. These materials are being used in urban areas to prevent flooding and manage rainwater more effectively, aligning urban development with sustainable water practices.

In terms of global engagement, the Netherlands continues to expand its role as a mediator and collaborator in international water conflicts and negotiations. Leveraging its extensive experience in managing complex water systems, the Netherlands offers mediation and technical assistance to countries experiencing transboundary water issues, promoting cooperation over competition. This diplomatic engagement not only helps resolve conflicts but also builds long-term capacity for joint water management among neighboring nations.

The future vision for Dutch water management also includes expanding the scope of their influence through digital platforms and tools. The Dutch water sector is developing sophisticated simulation software and decision-support systems that can be used by other countries to plan and manage their water resources more effectively. By sharing these tools, the Netherlands facilitates global access to advanced technologies that can lead to better water management outcomes worldwide.

In academia and research, Dutch institutions continue to be at the forefront of water-related studies, pushing the boundaries of what is known about water science. Collaborations between universities, private companies, and government agencies foster an environment where theoretical research quickly translates into practical applications. These partnerships not only drive innovation within the Netherlands but also contribute to the global body of knowledge on water management.

Through these comprehensive efforts, the Netherlands not only addresses its domestic water management challenges but also strengthens its leadership on the global stage. The Dutch model, characterized by its holistic approach and commitment to innovation and collaboration, serves as a guide for other nations aiming to enhance their resilience in the face of an increasingly uncertain water future. As global water challenges grow in complexity, the leadership and vision provided by the Netherlands will undoubtedly play a pivotal role in shaping sustainable water management practices worldwide.

At the community level, Dutch initiatives emphasize the importance of local engagement and education in water management. Through community-based programs, residents are encouraged to participate in water conservation efforts, such as rainwater harvesting and sustainable landscaping. These programs not only help reduce the community's water footprint but also raise awareness about the importance of water conservation. By empowering individuals and local communities, the Netherlands fosters a culture of stewardship that is essential for the long-term sustainability of water resources.

As the Netherlands continues to lead by example in the field of water management, it remains committed to the principles of innovation, sustainability, and collaboration. By developing and sharing cutting-edge technologies, advocating for robust global water policies, and fostering community involvement, the Netherlands not only addresses its own water challenges but also contributes to solving some of the most pressing water issues facing the global community today. This holistic approach ensures that the Dutch influence in water management will continue to be felt worldwide, driving progress toward a more sustainable and resilient future for all.

In conclusion, the Netherlands stands as a paragon of water management, skillfully navigating the complexities of both modern challenges and opportunities. Through a blend of advanced technology, proactive policy, and robust international collaboration, this small yet influential nation continues to lead the way in pioneering sustainable water solutions that cater not only to its unique geographical needs but also to the global community.

The Dutch commitment to continuous innovation in water management practices demonstrates an unwavering resolve to adapt to the evolving demands of climate change, population growth, and technological advancement. Their integrated approach ensures that every aspect of water management—from infrastructure and technology to policy and education—is geared towards sustainability and resilience.

Furthermore, by fostering global cooperation and sharing their extensive expertise, the Netherlands catalyzes worldwide efforts to manage water resources more effectively. This leadership is crucial in a world where water security is increasingly linked to national security, economic stability, and ecological sustainability.

As we look to the future, the vision and strategies deployed by the Netherlands offer valuable lessons for other nations grappling with similar water-related challenges. The Dutch model not only provides a blueprint for achieving water resilience but also inspires a collaborative approach to solving one of the most pressing global issues of our time. Through ongoing research, innovation, and international partnership, the Netherlands is not just solving its water management problems but is also helping to secure a more sustainable and water-wise future for the entire planet.

9

Conclusion: Lessons for the World

The Netherlands' approach to living with water offers profound lessons for the world, particularly as many regions face increasing threats from climate change, including rising sea levels and more frequent and severe weather events. Here are some key takeaways from Dutch water management strategies that other nations can adapt and implement.

The Dutch philosophy of water management is not just about preventing water from entering spaces where it's not wanted but rather integrating water into the urban and rural landscapes in ways that add value. This includes multi-functional spaces like water plazas that serve as recreational areas in dry weather and as water storage or runoff areas during heavy rains. Such an integrated approach encourages a balance between development and natural water cycles, enhancing both the utility and beauty of landscapes.

The Netherlands demonstrates the importance of incorporating proactive adaptation into global climate change policy. By planning for future scenarios that take into account the worst projections of sea-level rise and climate variability, Dutch policies ensure that infrastructure and communities remain resilient against potential disasters. Proactive adaptation involves constant updates to building codes, infrastructure norms, and urban planning guidelines to incorporate the latest data and technology available in water management.

Innovation is at the heart of Dutch water management, with continuous investments in new technologies and methods. From constructing state-of-the-art flood barriers to developing water-resistant materials for infrastructure, the Netherlands shows how engineering and technology can provide solutions to even the most daunting environmental challenges. Other nations can learn from this focus on research and development, applying innovative technologies tailored to their specific environmental and geographical needs.

Public education and engagement are crucial components of the Dutch strategy. The Netherlands invests in educating its citizens about the risks and management of water from a young age. This public awareness is critical in fostering a culture of water consciousness and participation, which ensures broader support for water-related policies and projects. Other countries can replicate this model to enhance public understanding and involvement in sustainable water management practices.

The Netherlands' role in global water management extends beyond its borders through active participation in international forums, sharing best practices, and aiding in water management projects worldwide. This global leadership not only helps other nations develop their water management strategies but also solidifies international cooperation on pressing water issues. Other nations can look to engage similarly, both to benefit from shared knowledge and to contribute to global solutions.

The Dutch commitment to green infrastructure and nature-based solutions serves as a model for environmentally friendly water management. Techniques such as creating urban green spaces that absorb rainwater, restoring natural waterways, and using vegetation to strengthen flood defenses not only mitigate water-related risks but also enhance urban biodiversity and improve the quality of life for residents. These practices promote ecological balance and offer a template for other nations looking to integrate similar sustainable practices in their urban planning.

The emphasis on collaboration across sectors and disciplines is another vital aspect of the Dutch approach. By involving engineers, urban planners, policymakers, businesses, and the community in water management discussions, the Netherlands fosters a holistic understanding of water issues that leads to more comprehensive and effective solutions. This multidisciplinary approach is crucial in addressing the complex challenges posed by water management and can be adopted by other nations to enhance their own strategies.

The Dutch also recognize the importance of flexibility in water management policies, allowing them to respond dynamically to new information and technologies. This adaptability is crucial in a world where climate change introduces significant uncertainties in water availability and demand. Learning from this, other countries can develop water management frameworks that are both robust and flexible, capable of evolving with changing environmental conditions and advancements in technology.

The Dutch model of water management, characterized by its integration, innovation, and international collaboration, provides valuable lessons for the global community. As countries around the world grapple with increasing water management challenges, the principles and practices honed by the Netherlands offer a guide to developing resilient, sustainable, and adaptive water management strategies. By embracing these lessons, other nations can not only safeguard their water resources but also enhance their environmental and social landscapes, building a more sustainable future for all.

As the world continues to confront the escalating impacts of climate change, the need for resilient water management systems becomes more acute. The Netherlands, with its proactive and integrated approach, exemplifies how strategic planning and innovative technologies can mitigate risks and enhance the adaptive capacities of water systems.

The Dutch focus on comprehensive risk assessment and management strategies also offers a blueprint for disaster preparedness. Their systematic approach to evaluating vulnerabilities and potential impacts of water-related disasters ensures that both urban and rural areas are better equipped to handle emergencies. This methodical assessment helps in prioritizing interventions and directing resources where they are most needed, a practice that can be instrumental for other nations facing similar threats.

The Netherlands' investment in public-private partnerships (PPPs) to finance and manage water projects illustrates an effective model for leveraging the strengths of both sectors. These partnerships often lead to enhanced efficiency, innovation, and capital investment in water management projects. Other countries can learn from this approach to mobilize additional resources, share risks, and accelerate the implementation of water management solutions.

In addition to physical infrastructure and technological innovations, the Dutch also emphasize the role of community engagement and legal frameworks in water management. They understand that sustainable water management is not only about constructing barriers and upgrading systems but also about fostering a regulatory and community environment that supports long-term water sustainability. Legislation that enforces sustainable practices and community programs that encourage water stewardship are crucial in cultivating a culture that values and actively participates in water conservation.

The Netherlands serves as a leader in international water law, advocating for equitable and sustainable management of transboundary water resources. Their diplomatic efforts and expertise in negotiation facilitate cooperation among countries sharing water resources, helping to prevent conflicts and encourage sustainable usage. This diplomatic approach is particularly relevant for regions where water resources cross national boundaries and where cooperative management could significantly enhance regional stability and development.

Continuing to share these insights and innovations, the Netherlands not only addresses its challenges but also contributes to global water security. As nations around the world adapt to the increasing variability and scarcity of water resources, the lessons from the Dutch experience provide valuable guidance on creating more resilient, sustainable, and inclusive water management systems. These lessons underscore the importance of a coordinated global response to water challenges, emphasizing that water security is not solely a national issue but a global imperative that requires collective action and shared responsibility.

The global community can further benefit from the Dutch emphasis on data-driven decision-making in water management. The Netherlands has harnessed the power of advanced data analytics, remote sensing, and IoT technologies to monitor water systems in real-time, predict potential issues, and respond proactively. This high level of data integration aids in precise decision-making and enhances the efficiency and effectiveness of water management strategies. By adopting similar data-centric approaches, other nations can improve their capacity to manage water resources, especially in facing the challenges posed by climate variability and population growth.

The Netherlands' approach to education and capacity building in water management extends beyond its borders. Through various international programs and initiatives, Dutch institutions share their knowledge and expertise in water science and engineering, helping to build global capacities. These educational exchanges not only disseminate valuable technical knowledge but also foster international networks of water professionals who can collaborate on global water challenges.

The concept of adaptive management is another significant aspect of the Dutch strategy that is crucial for global adoption. Recognizing that water management is an evolving challenge, the Netherlands implements policies and projects that are designed to adapt over time. This flexibility allows for adjustments based on environmental changes, technological advancements, and socio-economic dynamics. Other countries can incorporate adaptive management principles into their water policies, ensuring that they remain relevant and effective as conditions change.

The Dutch also recognize the importance of ecological and environmental sustainability in water management. Their projects often incorporate ecological restoration as a core component, aiming to enhance natural water retention and purification processes. By mimicking or supporting natural systems, these projects not only tackle water management issues but also contribute to biodiversity conservation and ecosystem health. This holistic view of ecosystem-based management is a critical lesson for the world, highlighting the interconnectedness of water management with broader environmental and sustainability goals.

As leaders in global water management, the Netherlands advocates for international collaboration and solidarity in addressing water issues. Through platforms like the United Nations and various international water management forums, the Dutch push for policies and practices that ensure water equity and sustainability worldwide. Their leadership in international discussions and negotiations underscores the global nature of water challenges and the need for united efforts in addressing them.

The Netherlands not only continues to refine its own water management practices but also actively contributes to shaping a more water-resilient world. The lessons drawn from Dutch experiences offer a comprehensive roadmap for other countries to enhance their water management systems, ensuring that communities worldwide can thrive in an increasingly water-variable future.

Building on its foundation of global cooperation, the Netherlands advocates for a unified approach to address the water crises affecting vulnerable populations worldwide. This commitment is demonstrated in their leading role in initiatives such as the World Water Forums and active participation in shaping the agendas of international water organizations. By pushing for policies that integrate water security with sustainable development goals, the Netherlands helps ensure that water remains at the forefront of international priorities.

The Dutch also leverage their advanced technological landscape to develop scalable solutions that can be implemented globally. Innovations such as water-efficient agricultural practices, smart irrigation systems, and climate-resilient infrastructure design are shared through international partnerships and development aid programs. These technologies are not only pivotal in addressing water scarcity but also in adapting to the changing climate, offering models that can be tailored to the specific needs and conditions of different regions around the world.

The Netherlands emphasizes the importance of resilient urban planning to manage water sustainably in city environments. With a significant portion of the global population living in urban areas, Dutch strategies for integrating water management into urban design provide a blueprint for cities worldwide. Techniques like creating flood-resilient public spaces, enhancing urban greenery, and implementing sustainable drainage systems are part of a broader approach to make cities more livable and climate-adaptive.

The proactive stance of the Netherlands also extends to financing water management. Recognizing the need for substantial investment in water infrastructure, the Dutch government and private sector collaborate on funding models that can be replicated internationally. These models often combine government subsidies, private investment, and international financing tools to fund large-scale water projects. By sharing their experiences in successful funding strategies, the Netherlands assists other nations in overcoming financial barriers to implementing comprehensive water management solutions.

The Netherlands is a proponent of continuous learning and innovation in water management. Dutch research institutions and companies are at the cutting edge of developing new water purification technologies, flood defense mechanisms, and sustainable water use practices. The commitment to research and development ensures ongoing improvement in water management techniques, setting a standard for academic and practical advancements in the field.

As countries worldwide grapple with the complexities of managing their water resources amidst climate change and urbanization, the Dutch model offers a multifaceted approach that emphasizes innovation, collaboration, and sustainability. The comprehensive strategies developed by the Netherlands serve not only their national interests but also as a guide for global water management efforts, aiming to achieve a water-secure future for all.

In conclusion, the Netherlands' approach to water management serves as a comprehensive model that encapsulates innovation, sustainability, and global leadership. With a rich history of contending with water-related challenges, the Netherlands has transformed its expertise into a suite of strategies that are not only effective domestically but also highly relevant for countries around the world facing similar issues.

The key lessons from the Dutch experience emphasize the importance of proactive adaptation, the integration of advanced technologies, and the value of public engagement and education in crafting effective water management policies. Moreover, the Netherlands' commitment to international collaboration and aid underscores the global nature of water issues and the necessity of collective action to address them.

By leveraging their knowledge and resources, the Netherlands contributes significantly to global efforts aimed at enhancing water security and resilience. This leadership is vital in an era where water management is increasingly recognized as central to addressing broader environmental and socio-economic challenges posed by climate change.

For other nations, the Dutch model offers valuable insights into developing resilient, adaptable, and integrated water management systems. It highlights that with thoughtful planning, innovative solutions, and international cooperation, it is possible to turn the challenges of water management into opportunities for enhancing societal well-being and environmental sustainability.

As the world continues to navigate the complexities of global water management, the lessons from the Netherlands will undoubtedly remain relevant, guiding efforts to ensure a sustainable and water-secure future for all. Through ongoing innovation and collaboration, the Dutch approach not only addresses the needs of the present but also lays a robust foundation for meeting the challenges of tomorrow.